Gift of the
John and Peggy Maximus Fund

FUTURISTIC ELECTRIC Trucks

Kerrily Sapet

PUBLISHERS

2001 SW 31st Avenue
Hallandale, FL 33009
www.mitchelllane.com

Copyright © 2020 by Mitchell Lane Publishers. All rights reserved. No part of this book may be reproduced without written permission from the publisher. Printed and bound in the United States of America.

First Edition, 2020.
Author: Kerrily Sapet
Designer: Ed Morgan
Editor: Sharon F. Doorasamy

Series: Futuristic Electric
Title: Trucks / by Kerrily Sapet

Hallandale, FL : Mitchell Lane Publishers, [2020]

Library bound ISBN: 9781680203561
eBook ISBN: 9781680203578

PHOTO CREDITS: Design Elements, freepik.com, Shutterstock.com, Cover Photo: JoachimKohlerBremen CC-BY-SA-4.0 , p. 5 Martin Dürrschnabel CC-BY-3.0, p. 6 loc.gov, p. 9 Korbitr imgur, p. 11 Linneakornehed Wikipedia CC-BY-SA-4.0, p. 12 Theurv Wikipedia CC-BY-SA-4.0, p. 17 Toll Group CC-BY-2.0, p.19 USMC public domain, p. 25 Mr.choppers Wikipedia CC BY-SA 3.0, p. 27 Brian Snelson CC-BY-2.0

CONTENTS

Chapter One
FROM SLEDS TO SEMIS — 4

Chapter Two
GOING ELECTRIC — 8

Chapter Three
MOTORS IN MOTION — 14

Chapter Four
GREEN TRUCKS — 20

Chapter Five
A CLEAN FUTURE — 24

WHAT YOU SHOULD KNOW — 28
GLOSSARY — 29
WORKS CONSULTED — 30
FURTHER READING — 31
ON THE INTERNET — 31
INDEX — 32
ABOUT THE AUTHOR — 32

Words in **bold** throughout can be found in the Glossary.

Chapter One

FROM SLEDS TO SEMIS

Long ago, when roads were just tracks formed by footsteps, people moved heavy objects by dragging them on wooden sleds. Someone in Mesopotamia 5,000 years ago flipped two pottery-making wheels on their sides and fastened them to a sled to move faster. The idea spread, and people began attaching wheels made of stone or tree trunks to create carts and wagons. Soon, roads stretched like spider webs between cities. Wagons pulled by oxen, yaks, and horses rolled along the world's longest road—the Silk Road—from Africa to China. They carried silk, gold, spices, and inventions.

Inventors tried to create engines to do the work of animals, in order to travel faster and haul goods more easily. By the 1600s, people were experimenting with large, clunky carriages powered by steam engines. Soon they were testing motors powered by electricity and engines powered by gas.

German inventor Gottlieb Daimler built one of the first gas engines strong enough to move a vehicle. Daimler had always loved designing and tinkering with machines. As a child, he took technical drawing classes. As an adult, he worked in a steam engine factory. Daimler used his backyard greenhouse as a workshop. Neighbors thought he was printing counterfeit money. When the police showed up, they discovered that he was building engines.

Daimler attached his engine, called the Phoenix, to the first car with four wheels. In 1896, he built the first truck—a carriage with iron-clad wooden wheels. Daimler discovered his "motorized goods vehicle" was difficult to sell. It cost more than a horse and wagon, and gasoline was only sold at pharmacies as a cleaning product.

This 1898 Daimler Motor-Lastwagon, the world's oldest surviving goods vehicle, is on display at the Mercedes-Benz museum in Stuttgart, Germany.

CHAPTER ONE

American Express trucks transporting gold shipped by Great Britain to New York City for safekeeping during World War I, in 1915.

Cars and trucks began to replace horse-drawn wagons and carriages as people realized that vehicles with gas engines were faster, lighter, and more powerful. They also didn't get thirsty, hungry, or sick like horses. When World War I broke out in 1914, the truck business boomed. Armies wanted tough, rugged trucks. Engineers began building bigger trucks with stronger engines and tires that gripped the road.

Today, trucks come in all sizes—from pickup trucks to tractor-trailers (also called semis or 18-wheelers). More than 200 million trucks travel the roads. They carry everything from pigs to popsicles, picking up **freight** from ports and railroad yards and delivering it to stores and homes. They crisscross every continent, driving across bridges, through deserts, over mountains, and even frozen lakes.

From Sleds to Semis

Most trucks run on diesel, a fuel made from petroleum. Petroleum is a fossil fuel, formed over millions of years from layers of rock crushing the remains of plants and animals. Fossil fuels are a **nonrenewable resource**: when they're used up, they're gone. When fossil fuels are burned, they produce carbon dioxide, which builds up in the Earth's **atmosphere** and causes climate change. They also release **particulates**, tiny pieces of leftover fuel, into the air. Particulates contain chemicals that harm people and animals.

Air pollution causes breathing problems and contributes to millions of deaths every year. Many people think trucks powered by electricity instead of fossil fuels are the solution to pollution.

Fun Facts

1. Every year, trucks in the United States drive 450 billion miles—driving the distance from Earth to Saturn and back 300 times.

2. Trucks produce billions of tons of carbon dioxide every year.

Chapter Two

GOING ELECTRIC

In the early 1900s, some electric delivery trucks puttered down the streets delivering milk, eggs, and bread. Trucks with gas engines were speedier and could carry heavier loads. They soon replaced electric trucks. Today, electric trucks are making a comeback. People liked them 100 years ago for the same reasons they do now.

Electric trucks are quieter and less expensive to operate. They run on electric motors powered by rechargeable batteries instead of engines powered by fossil fuels. The motor turns the truck's wheels. Electric trucks also are **zero-emission** vehicles, meaning they don't produce any pollution.

Major truck manufacturers, such as Daimler, Volvo, BYD, and smaller companies are racing to build electric trucks. Trucking companies "don't want science projects," says Giordano Sordoni, cofounder of Thor Trucks. "Electric addresses the cost of fuel and maintenance all in one fell swoop."

Tesla's electric truck, the Tesla Semi, hauls about 44,000 pounds and goes 500 miles. "[It's] worth seeing this beast in person," says Elon Musk, Tesla's cofounder. "It's unreal." Using Tesla's Megachargers, the Semi recharges in 30 minutes. "Because these Megachargers are solar powered, your truck will be running on sunlight," Musk says. Tesla plans to start producing the Semi in 2019.

A Tesla Semi, an all-electric semitrailer with an estimated range of about 500 miles on a full charge

CHAPTER TWO

Thor Trucks' electric truck, the ET-One, can go 300 miles before recharging. Dakota Semler, the company's cofounder, has been building green cars for years. As a teenager, he converted his mother's SUV to run on vegetable oil. The ET-One looks like the helmet of a medieval knight. "Everyone thinks we've built a Transformer," says Semler. The ET-One will hit the streets by 2019.

Daimler Trucks is designing several electric trucks. The Mercedes-Benz eActros, being tested in Germany and Switzerland, is a city delivery truck that will be operational by 2020. The Freightliner eCascadia, a massive tractor-trailer, is due out by 2021.

Not all new designs run on batteries. Nikola Motor Company's electric truck runs on fuel cells, which combine hydrogen and oxygen to produce electrical energy. The company's truck, the Nikola One, has a range of 1,200 miles. They call it the "iPhone of trucking" and plan to have it running by 2021.

Other companies are testing self-driving trucks. Einride's T-Pod is a driverless electric delivery truck. The windowless, seatless truck looks like a giant freezer on wheels and uses sensors and cameras to travel along its route.

Going Electric

"This invention makes today's truck as outdated as horse and carriage."

—ROBERT FALCK, EINRIDE

Einride's T-pod, a full-scale heavy-duty truck that can be operated remotely by a driver miles away

CHAPTER TWO

Many companies, such as UPS, are testing electric trucks in cities in the United States, Europe, and China. Some are already driving down the streets. Others are coming within the next few years.

Fun Fact

In 2014, Chicago became the first city in the United States to use electric garbage trucks.

Going Electric

10-4, GOOD BUDDY

USING CB RADIOS, TRUCKERS TALK TO EACH OTHER AND SHARE NEWS USING THEIR OWN LANGUAGE. A TRUCKER DRIVING TO "MICKEY MOUSE" (ORLANDO, FLORIDA) CARRYING A LOAD OF TOOTHPICKS (LUMBER) MIGHT EAT AT A "CHOKE AND PUKE" (A RESTAURANT WITH BAD FOOD), FILL UP WITH "GO-GO JUICE" (GASOLINE), AND KEEP THE "SHINY SIDE UP" (NOT FLIP OVER).

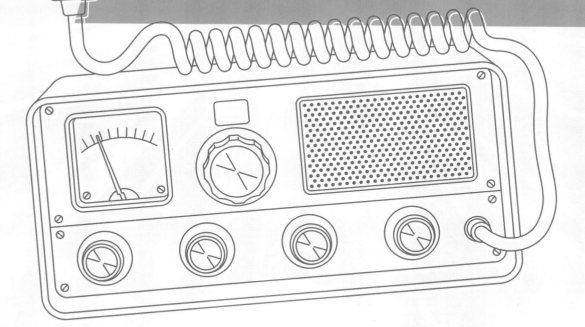

Chapter Three

MOTORS IN MOTION

Engineers designing electric vehicles think about motors and batteries instead of engines and diesel. But building an electric truck isn't as easy as swapping out an engine for an electric motor. Electric motors work differently than gas engines.

Gas engines squeeze fuel and oxygen to create small explosions that push metal parts connected to a **driveshaft**. The driveshaft turns the wheels. In an electric motor, the motor and the driveshaft don't touch. The driveshaft is pushed magnetically. Electric motors are "a very beautiful, efficient thing," says Dustin Grace, an engineer.

Electric motors are like light switches. They turn on instantly and produce power. Electric motors produce instant **torque**, the force that turns the wheels and helps tow weight. Electric vehicles are more powerful at lower speeds and better at climbing hills.

Electric motors have one moving part. Gas engines have hundreds of parts swimming in oil to keep them moving smoothly. Because electric motors have fewer parts, they are easier to repair and need less maintenance. Electric motors also feed themselves electricity. When a driver puts on the brakes, the motor runs in reverse, sending electricity back to the battery.

Electric motors are smaller than gas engines. A motor the size of a beach ball powers Thor Trucks' ET-One. Engineers design electric trucks differently because they don't need space for engines the size of washing machines. With no engine under the hood, the front of the Tesla Semi slopes down and has a "frunk," a front trunk for storage.

An artist's depiction of an electric motor

CHAPTER THREE

Engineers create **aerodynamic** designs so air flows smoothly around the truck and the motor doesn't fight the wind to move the truck forward. They test their designs in wind tunnels and use computers to analyze the results. "We designed the Tesla truck like a bullet," says Elon Musk of Tesla. "A normal diesel truck is designed like a barn wall."

Electric motors run on electricity stored in lithium-ion batteries—the same type of batteries that power cell phones and computers. Electricity is measured in **watts**. A kilowatt-hour is the power supplied by 1,000 watts for one hour. Leaving a television on for four hours uses one kilowatt-hour. The more electricity a battery stores, the farther a vehicle goes before it needs to recharge. The Freightliner eCascadia's 550 kilowatt-hour batteries allow it to go 250 miles.

Batteries don't match the power of fossil fuels. A battery needs to weigh 100 pounds to equal the driving distance of one pound of a fossil fuel. Tractor-trailers hold 1,000 pounds of fuel. To balance the extra battery weight, engineers place the battery packs under the cargo area or along the sides of the truck's frame. The axles and wheels also help spread out the weight.

Motors in Motion

Toll's 10-ton electric truck by Smith Electric Vehicles

CHAPTER THREE

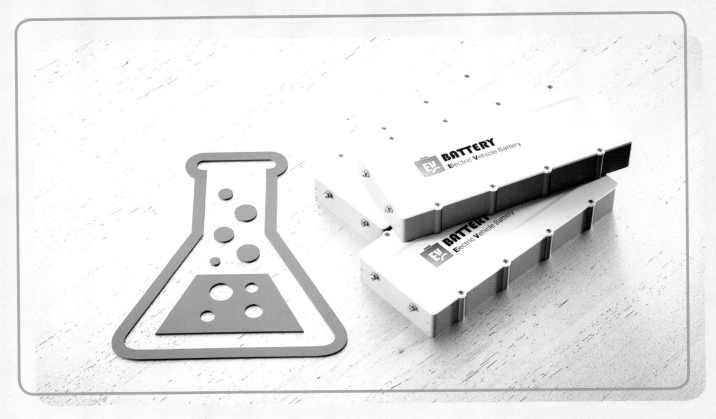

The heavier the load, the harder the motor works and the faster the battery drains. Battery scientists are testing thousands of chemicals to invent lighter, stronger, longer-lasting batteries. Batteries also don't work as well in the cold. Scientists are testing electric trucks fitted with battery heaters in Antarctica to learn more about electric vehicles in harsh weather.

Engineers also are building trucks with **composites**, such as carbon fiber. Composites are lightweight, strong materials made by combining two different materials. Electric trucks of the future will feature advanced technology. The new designs look like powerful race cars built to haul the weight of four elephants. "Everything people know about trucking is going to be changed," says Trever Milton of Nikola Motor Company.

Fun Facts

1. BYD, a Chinese company, is the world's largest manufacturer of electric vehicles.

2. The Shockwave, the world's fastest truck, has three jet engines and travels 370 mph.

Chapter Four

GREEN TRUCKS

Designing, building, and testing a new truck takes time and money. Today, companies around the world are spending billions of dollars to develop electric trucks. These new designs, called **prototypes**, range in size from pickup trucks to tractor-trailers. "The future of electric transportation is happening now," says Dakota Semler of Thor Trucks.

Thor Trucks, a California company, is working on the ET-One, a delivery truck that can go 70 mph and recharge in 90 minutes. Some call the ET-One a "Frankenstein of trucks" because its frame, motor, and axles are made by different companies. The ET-One is due out by 2019.

Two Tesla Semis

The Tesla Semi could also be on the streets by 2019. Four electric motors drive the truck's rear wheels. "Even if two of the motors quit, there are two more, and it can still beat a diesel," says Elon Musk of Tesla. Although the truck doesn't "transform into a robot and fight aliens," like Musk joked, it features dashboard touch-screens and cameras instead of side mirrors.

Daimler Trucks has several prototypes. The Mercedes-Benz eActros can make deliveries during the day and recharge at night. With two electric motors by the rear wheels, it travels 125 miles before needing to recharge. The eActros will be out by 2020. Another prototype, the Freightliner eCascadia, is a heavy-duty electric truck capable of hauling 44,000 pounds.

CHAPTER FOUR

Some companies are designing pickup trucks. Workhorse's hybrid pickup uses battery power for the first 80 miles before a gas engine kicks in to power the truck.

> *"For the first time in 108 years, someone has invented a truck that's cheaper than a gasoline truck over its lifetime."*
>
> —STEVE BURNS, WORKHORSE

Tesla's electric pickup will be "a pickup truck that can carry a pickup truck," says Musk.

Some people predict self-driving trucks are the trucks of the future. Truckers spend hours at a time going straight on roads and often drive on empty highways. They travel at steady speeds and don't face pedestrian traffic. By law, truckers can only spend 11 hours at a time behind the wheel before resting. A self-driving truck can be programmed to stay in its lane and keep a safe distance. "It never gets tired. It's always 100% sharp. It's never angry. It's never distracted," says Wolfgang Bernhard of Daimler.

Einride's self-driving T-Pod trucks go 124 miles before recharging. The box-shaped trucks use sensors to drive on the highways. Humans using remote controls take over the trucks on main roads. Robert Falck of Einride says drivers can sit, "with a cup of coffee in hand, in front of a computer, driving and operating a fleet of T-Pods." Einride plans to have

Green Trucks

T-Pods on the roads in Sweden by 2020. Mercedes-Benz is also testing a self-driving truck. The company Embark is already using self-driving trucks to deliver refrigerators from Texas to California.

New environmentally friendly truck prototypes are starting to hit the streets and highways. Some trucks can even drive themselves. "People don't want noise, they don't want pollution, and pressure for change is coming," says Marc Llistosella of Mitsubishi Fuso. "We must take innovation by the horns and go." The future of trucking is changing fast.

Fun Fact

The solar-powered aCar, which is part pickup truck, part car, helps people in rural Africa travel rocky, hilly, unpaved roads.

LEGENDS OF LIGHTNING

ENGINEERS ARE OFTEN INSPIRED BY HEROES. NIKOLA MOTOR COMPANY AND TESLA ARE NAMED AFTER NIKOLA TESLA, AN INVENTOR WHO STUDIED ELECTRICITY. THOR TRUCK'S NAME AND LOGO ARE INSPIRED BY THOR, THE HAMMER-WIELDING NORSE GOD OF THUNDER AND LIGHTNING. EINRIDE'S RAM'S HEAD LOGO ALSO CELEBRATES THOR, WHO RODE ON A CHARIOT PULLED BY TWO RAMS.

Chapter Five

A CLEAN FUTURE

Close your eyes and imagine a highway without trucks roaring past spewing thick clouds of exhaust. Picture a city without the black haze of air pollution with clean, quiet trucks making deliveries. People around the world have those same visions as they design and build electric trucks. Electric trucks are driving down roads—whether made of asphalt, dirt, or ice—on every continent from Africa to Antarctica.

Electric trucks in the early 1900s ambled down streets and only traveled about 50 miles. With advances in motors, batteries, and materials, today's electric trucks can go six times faster and travel hundreds of miles. The midsize electric trucks driving down streets today make deliveries around cities and drive short routes before returning to the depot at night to recharge their batteries.

Delivery companies, such as UPS, are converting their fleet to electric trucks. UPS is testing trucks designed by Workhorse and Daimler. Electric trucks are "perfect for large, heavily populated urban centers," says Bryan Allen of Daimler. UPS plans to use 1,500 electric trucks in New York City by 2020. "It's the beginning of the end" for gas-powered engines, UPS says.

An artist's depiction of an electric delivery van at a charging station

CHAPTER FIVE

Many countries, such as China, France, India, and the United Kingdom, and several cities in the United States, have proposed bans on engines that burn fossil fuels. The company BYD predicts that all of China's trucks will be electric by 2025. Electric trucks are cleaner and quieter. They also need fewer repairs and are cheaper to operate.

"Electric truck technology is not just a way to make money—it's good for the environment and good for people," says Giordano Sordoni of Thor Trucks.

In 2017, the number of electric trucks in the world increased from 40,000 to 371,000. Experts predict that number will keep increasing. Although long-haul electric tractor-trailers have yet to drive down the highway, the race is on to build them. Self-driving trucks may even be making deliveries within the next few years. "If I brought out my crystal ball, I'd say sooner than you think," says Roger Nielsen of Daimler Trucks.

Electric trucks can haul everything from flowers and furniture to eggs and electronics, without polluting cities and contributing to climate change. They are good for people and for the planet. Electric trucks are hitting the streets in cities around the world. One may even be delivering packages in your neighborhood soon.

A Clean Future

Fun Facts

1 Australian "road trains" have three or more trailers and are equipped with bars to protect trucks from collisions with kangaroos.

2 Truckers in India paint their trucks with a kaleidoscope of colorful designs to attract business.

What You Should Know

- In 1896, Gottlieb Daimler builds the first gas-powered truck.
- In the 1900s, some companies use electric delivery trucks.
- In the 1920s, gas-powered trucks overtake electric trucks.
- In the 2000s, companies begin designing and testing electric trucks.
- In 2017, Tesla announces plans to build the Tesla Semi.

Glossary

aerodynamic
The way air moves around an object

atmosphere
The layer of gases surrounding a planet

composite
Made of various parts or materials

driveshaft
The parts of a vehicle that move the wheels

freight
Items carried by a truck, train, ship, or plane

nonrenewable resource
A natural supply that cannot be replaced after it is used

particulates
tiny pieces of leftover fuel, into the air

prototype
The first model of a new machine or design

torque
the force that turns the wheels and helps tow weight

watts
A measurement of electricity

zero-emission
An engine, motor or process without waste products that cause pollution

Works Consulted

Gies, Erica. "Electric Trucks Begin Reporting for Duty, Quietly and Without All the Fumes." *Inside Climate News*, December 18, 2017.

Hyatt, Kyle. "Swedish Startup Einride Prepares to Drop the T-Pod at Detroit." C-Net.com, January 3, 2018.

Korosec, Kirsten. "Nikola Motor Will Build its Electric Semi Trucks in Arizona." *Fortune*, January 30, 2018.

O'Dell, John. "Elon Musk Unveils Superfast 500-Mile Range Tesla Electric Semi Truck." Trucks.com, November 17, 2017.

"Power to the Truck—the History of Electrically Driven Commercial Vehicles." Daimler.com, accessed July 6, 2018.

Straight, Brian. "Electric Trucks are Coming Says DTNA President." FreightWaves.com, September 25, 2017.

Vance, Ashlee. "This Electric Truck Will Probably Beat Tesla's to Market." *Bloomberg*, December 13, 2017.

ZumMallen, Ryan. "Thor Trucks Storming into Heavy-Duty EV Market." Trucks.com, December 13, 2017.

Further Reading

Amstutz, L. J. *How Can We Reduce Transportation Pollution?* Minneapolis, MN: Lerner Publications Company, 2016.

Harris, Tim. *Transportation Technology.* New York: Cavendish Square, 2016.

Richards, John. *It'll Never Work: Cars, Trucks, and Trains: An Accidental History of Inventions.* New York: Franklin Watts, 2017.

On the Internet

http://amhistory.si.edu/onthemove/exhibition/
A website of the Smithsonian National Museum of American History

http://kids.saveonenergy.ca/en/index.html
An interactive webpage by the Independent Electricity System Operator

https://www.eia.gov/kids/
A kid-friendly website sponsored by the U.S. Energy Information Administration

Index

batteries	8, 10, 14, 15, 16, 18, 22, 25
Daimler	5, 9, 21, 22, 25, 26, 28
Einride	10, 11, 22, 23
fossil fuel	7, 8, 26
eCascadia	10, 16, 21
Musk, Elon	9, 16, 21, 22
Nikola Motor Company	10, 19, 23
pollution	7, 8, 24
Semler, Dakota	10, 20
Sordoni, Giordano	9, 26
Tesla Semi	9, 15, 21, 28
Thor Trucks	9, 10, 15, 20, 23, 26
T-Pod	10, 11, 22, 23
UPS	12, 25
Workhorse	22, 25

About the Author

KERRILY SAPET is the author of more than 20 nonfiction books and multiple magazine articles for kids. She has driven pickup trucks and moving trucks and traveled up the Pacific coast from Washington to California in a semitruck. Sapet currently lives near Chicago, Illinois, with her husband and two sons.